DECISION

OF

HON. PHILIP F. THOMAS,

COMMISSIONER OF PATENTS,

ON THE

APPLICATION OF SAMUEL F. B. MORSE,

FOR AN EXTENSION OF HIS PATENT FOR A NEW AND USEFUL IMPROVEMENT IN

ELECTRO-MAGNETIC TELEGRAPHS.

Patented April 11, 1846.—Patent Extended for Seven Years from the 11th Day of April, 1860.

WASHINGTON:
HENRY POLKINHORN, PRINTER.
1860.

DECISION

OF

HON. PHILIP F. THOMAS,

COMMISSIONER OF PATENTS,

ON THE

APPLICATION OF SAMUEL F. B. MORSE,

FOR AN EXTENSION OF HIS PATENT FOR A NEW AND USEFUL IMPROVEMENT IN

ELECTRO-MAGNETIC TELEGRAPHS.

PATENTED APRIL 11, 1846.—PATENT EXTENDED FOR SEVEN YEARS FROM THE 11TH DAY OF APRIL, 1860.

WASHINGTON:
HENRY POLKINHORN, PRINTER.
1860.

U. S. Patent Office, April 10, 1860.

IN THE MATTER OF THE APPLICATION OF SAMUEL F. B. MORSE FOR AN EXTENSION OF LETTERS PATENT GRANTED TO HIM ON THE 11TH OF APRIL, 1846, AND RE-ISSUED ON THE 13TH OF JUNE, 1848, FOR A NEW AND USEFUL IMPROVEMENT IN ELECTRO-MAGNETIC TELEGRAPHS.

In this case, the utility of the invention, its value and importance to the public, and the diligence employed in introducing it into general use, are not denied by the parties who oppose the grant of the extension.

It is, however, insisted that the prayer of the petitioner ought not to be granted, for two reasons: first, because the invention was not novel and patentable when the patent was issued; secondly, because, if the patentee was really the first and original inventor of the improvement covered by the patent, he has already received a reasonable remuneration for the time, ingenuity, and expense bestowed upon it, and upon its introduction into use.

Although the validity of the patent granted to Professor Morse in the year 1840, for an electro-magnetic telegraph, is not now a legitimate question for the consideration of this Office, the intimate relations subsisting between the two inventions, as well as the range which the argument on both sides of this application has been allowed to take, seem to render a reference to the character and scope of that patent indispensable to an intelligent discussion and understanding of the nature and extent of the claims included in, and protected by, the grant now sought to be extended.

Morse's invention, as described and claimed in the re-issue, dated June 13, 1848, of the patent of June 20, 1840, consisted in the application of galvanic electricity to telegraphic purposes, by a combination of electric currents and machinery. Its details embraced: first, a galvanic battery, a galvanic circuit, a type rule, and signal lever, combined with an electro-magnet, by means of which, the motive power of a current of electricity was developed and applied to give motion to other machinery, for the purpose of registering permanent characters at any distance; second, a register, consisting of a pen lever, with a marking device at one extremity, and an armature, to be acted on by an electro-magnet at the other, in combination with a roller, operated by a clock movement, over which a continuous ribbon of paper was drawn, receiving, in its passage under the pen lever, the message, in permanent dots and lines, with spaces between them.

The claim, under this re-issue, embraced several of the mechanical devices employed, together with combinations of two or more galvanic circuits, with independent batteries.

At the period when Morse was engaged in his efforts to make and perfect his invention, the opinion was prevalent among men of science, whose attention had been attracted to the subject, that the electric current could not be transmitted through a circuit of any considerable distance, with sufficient force to imprint intelligible signs or characters at its terminus. Hence, was suggested the idea of a combination of two or more shorter circuits, each having its own battery, coupled together and forming links or relays in the main line, for the purpose, by means of breaking and closing the successive circuits, of renewing the galvanic impulse, and propelling forward the electric current, with sufficient power to work a register at the end of the line, and at an indefinite distance from the initial point, and, thereby, to transmit and record messages. This combination of circuits was the main feature in Morse's first invention, and, with the mechanism above referred to, the alphabet, and other less important devices, formed the subject-matter of his patent of the 20th of June, 1840, re-issued on the 13th of January, 1846; again re-issued on the 13th of June, 1848, and extended for the term of seven years from the 20th of June, 1854, the date of its expiration.

The title of Professor Morse to be considered and treated as the first and original inventor of the arrangements and devices protected by this patent, is not, now, an open question.

On four several occasions, his specifications and claims have been submitted to the scrutiny of this Office, and whatever of merit attaches to the invention, has been, as often, awarded to him. The same patent, with like results, has passed the ordeal of the federal courts in Massachusetts, Pennsylvania, and Kentucky; and its validity, with the exception of the eighth clause of the claim, which was very broad, has been vindicated by the solemn judgment of the Supreme Court of the United States. It is, therefore, no longer within the control of the Patent Office. But, still, it forms the basis of important improvements, made by the same inventor, in the telegraphic art; for, notwithstanding the novelty and utility of that invention, it was, in its practical operations, very far from attaining the degree of excellence to which subsequent improvements have brought it. In its practical results, two very marked defects were apparent, which, in point both of convenience and economy, rendered it extremely inefficient.

In the first place, the receiving magnet would work only in one direction, and, therefore, although messages might be sent from the initial point, they could not be returned from the opposite end of the line, over the same wire. Between the termini of the main line, despatches could not be interchanged. Under such circumstances, two distinct and independent metallic circuits, each requiring to be kept constantly in working order, were indispensable to its successful operation. Then, again, communication could only be had through the relay circuits, with the terminal points of the line, there being no provision for the delivery of despatches at intermediate or lateral stations, without interruption to the main line, except through the agency, also, of separate and independent circuits.

The invention patented on the 11th of April, 1846, reissued on the 13th of June, 1848, and now sought to be extended, was designed to obviate and remove these defects. Experiments made by Professor Morse subsequently to the date of the patent of 1840, demonstrated that the theory on which the invention of the combined or relay circuits was founded, was erroneous; that magnetic power sufficient to move a pen lever, could be obtained from the electric current on a circuit of indefinite length, and that there was no necessity for combining two or more circuits, for the purpose of renewing the impulse, at short intervals, on the main line. In the new invention, therefore, the combined circuits, with the relay magnets, were abandoned, and, in their stead, was substituted a single wire, capable of being worked both ways, connected by receiving magnets with short lateral or office circuits, upon which the requisite registers were placed. Thus, the chief defects in the first invention were effectually removed. Messages could be transmitted on a single wire, in both directions, between the termini of the line, and simultaneously delivered, through the lateral circuits, at way stations, without interruption to the main or local lines, and without impairing, for a moment, the efficiency of either.

The invention now under consideration, may be said to involve the employment of two or more single circuits, consecutively arranged, each having a battery and receiving magnet, or their equivalents, and the combination of one main single circuit, and any number of receiving magnets, each receiving magnet closing and breaking an independent circuit, which forms no part of the main line; nor is the main line influenced in its action by the derangement of any one, or all, of the local or secondary circuits. Thus, the main single circuit is made to connect with one, or any number of receiving magnets, and bring into action its local battery, to operate the armature of the pen lever, and permanently record the message on paper. It may, therefore, be proper to regard the leading claim in this invention, as consisting in the combination of a main circuit with a receiving magnet and a local battery, at any number of stations between the termini of the main circuits, where the local battery is arranged to operate the mechanism employed in recording the message. Morse's own description of the invention, furnished in his testimony in the suit of French *vs.* Rogers, made, by consent of parties, evidence in this case, will, perhaps, convey to the mind of the superficial reader, a more intelligible idea of its character than can be elsewhere found. It is as follows:

"In its present practical form, it consists of the following 'parts: a main, or long circuit of wire, which is extended 'between any two or more distant places, and through which 'main circuit the electric current from any battery or generator 'of electricity is made to act at pleasure, through two helices 'around two cores of soft iron, making an electro-magnet, 'which, attracting an armature attached to a lever or its 'equivalent, produces motion in such lever or its equivalent, 'either to act directly in marking upon paper or other suit'able material, at practicable distances, or, indirectly, by 'touching off, so to speak, a second or short circuit, and as 'many local circuits along the line as may be desired, each 'containing a battery, which imparts power sufficient to ac'complish the great result, to wit: to mark or imprint char'acters at a distance, and to do this at as many places along 'the line, at the same time, as may be desired. The resist'ance to be overcome, is the slight *vis inertiæ* of a delicate 'armature upon a balanced lever and a delicate spring. The 'lever breaks and closes by the power thus exerted by the 'secondary or short circuit. The short circuit operates di'rectly upon the register, which is an instrument of clock'work for carrying paper, so that the paper, while in motion, 'is marked on by a pen or stile in contact with it, and thus 'forming on it, the dots and lines specially and peculiarly 'adapted and applied to the purpose."

Such is the invention, the patent for which, this Office is now asked to extend for an additional term of seven years.

It will be perceived that the main claim under this patent consists of a combination of devices, all of which were old; some of them, originating with Morse himself; some, due to the ingenuity of other inventors; but, all well known and in use prior to the grant of the patent of 1846.

In this state of the case, it is objected, on the part of the remonstrants, that the arrangement of the local circuit and local magnet was included in Morse's patent of 1840, and that, inasmuch as a second patent for the same invention cannot lawfully issue, either to the same grantee or to any other party, the patent of 1846 is a nullity, and ought not, therefore, to be extended.

It is a well established and, certainly, at this day, an admitted principle of patent law, that a combination of old devices producing new and useful results, is patentable.

In Ryan *vs.* Goodwin, 3d Sumner's Reports, 514, 518, a case in which the patent claimed, as the invention, a new and useful improvement in making friction matches, Justice Story remarked: "It is certainly not necessary that every ingredient, or, indeed, that any one ingredient used by the 'patentee in his invention, should be new or unused before 'for the purpose of making matches. The true question is, 'whether the combination of materials by the patentee is, 'substantially, new. Each of these ingredients may have 'been in the most extensive and common use, and some of 'them may have been used for matches, or combined with 'other materials for other purposes. But if they have never 'been combined together in the manner stated in the patent, 'but the combination is new, then, I take it, the invention of 'the combination is patentable."

The same doctrine has been adopted and repeatedly asserted in the practice of this Office.

In Ex parte Whittlesey, vol. 1, page 23, Commissioner's Decisions, the Commissioner said: "I fully assent to the 'proposition that a combination of old devices is patentable 'when a new and useful result is thereby attained, but, in 'such cases, something more than the mere assembling 'together of the several devices and placing them in juxta'position is requisite. They must be made to act in concert, 'in order to produce a result by their united action."

Again, in Wycoff's case, vol. 1, page 69, Commissioner's Decisions, it is said: "It is no objection to a patent for a 'new combination, that all the devices are old, or are the 'subject of other patents. And it matters not that the devices 'have before been grouped into minor combinations, and 'patented as such. In such cases, the new patentee cannot 'use the devices or combinations previously patented, with'out an arrangement with the several patentees of those 'parts of his machine. But, as against all others, except the 'patentees, his rights are complete."

And, again, in Denning's case, ibid, 128, "The applicant 'only claims a combination of devices, and whether they 'are new or old is wholly unimportant."

The current of decisions in the English courts runs in the same direction. In Huddart *vs.* Grimshaw, Webster's Patent Cases, 86, Lord Ellenborough remarked: "There are common 'elementary materials to work with in machinery, but it is 'the adaptation of these materials to any particular purpose 'that constitutes the invention; and if the application of 'them be new, if the combination in its nature be essentially 'new, if it be productive of a new end, and beneficial to 'the public, it is that species of invention which, protected 'by the King's patent, ought to continue to the person the 'sole right of vending."

The Supreme Court of the United States has, on several occasions, recognized the same doctrine as sound law, and in the case of O'Reilly et al. *vs.* Morse et al., 15th Howard, where the validity of this very patent was considered and decided, the Chief Justice, in delivering the opinion of the Court, said, in regard to the combination: "The elements 'which compose it may all have been used in the former 'invention; but it is evident that their arrangement and com-'bination must be different to produce this new effect." And, hence, the patent was adjudged to be valid.

Assuming, then, the result or effect produced by the combination in question to have been novel and useful, at the date of its first invention, a proposition which is not denied, the sole inquiry to which this office is limited, is, not whether the several elements composing it are old or new; not whether Morse was or was not the original inventor of all or any of them; nor whether, separately considered, they or either of them were included in his invention of 1840, or in that of any other inventor; but whether the same or equivalent devices, producing the same or like results, are to be found, in combination, in the patent of 1840, or in any other patent, or in any printed publication, or in public use prior to the invention covered by the patent of 1846.

The most reliable, and, indeed, the only test of the extent and scope of a patented invention, is the record of the patent itself, including, of course, the model, drawings, specifications, and claims. The record of the patent of 1840 has been already resorted to for the purpose of ascertaining, so far as relates to the subject of the present inquiry, the nature and character of the particular devices therein contained, and the results intended to be accomplished by them, or which, together, they were capable of producing, and from the description of them, hereinbefore given, it may be confidently asserted that nowhere in all that record can there be discovered a device, or combination of devices, by means of which the electric current, as it passes on the main line, and, without interruption in its course, is made to break and close local or lateral and independent circuits, whereby messages are recorded simultaneously at the end of the line, and at any number of local offices, at intermediate stations, or transmitted to any distance in a lateral direction. So far from this being the case, as has been already shown, the chief embarrassment experienced in operating the telegraph under Morse's first system consisted in its utter inability to record despatches at way stations without arresting the passage of the current on the main line, and, thereby, suspending all communication with its terminus; and the only means of obviating this difficulty, known at the date of the first patent, was the establishment of independent lines between the initial and lateral points. True, the local circuit at the end of the line, so regulated in its length as to cause the message to be effectually recorded, was included in the first invention, but it differed in nothing from the relay circuits in the same invention, save in its connection with the mechanism designed to operate the register. It is equally true, that the local circuit, with some of the recording devices, and other and different elements, has been, in the last combination, transferred to the margin of the line, to effect a novel and important result, but it is also true, that without an entire re-arrangement of some, and a change in other of its parts, its functions, in the new connection, could not have been performed, nor could the results intended by its transfer have been attained.

The objection, then, on this ground, cannot be maintained, for it is absolutely certain that the record of the patent of 1840 discloses no such combination as that specified and described in the patent of 1846.

But, it is contended, that if the combination of the local circuit and local magnet was not included in the record of the first patent, it was, in fact, invented and perfected by Morse in the winter of 1837-8, and, being an essential feature of his telegraphic invention of that year, subsequently patented in 1840, it was incumbent on him, under the law, to have described and claimed it in the specifications of that patent; that, failing to do so, the invention could not be protected by a separate grant, and that the patent of 1846 is, therefore, void, and ought not to be extended.

If this objection be well founded, it may be considered fatal to the validity, not only of the patent now under consideration, but, also, of that granted in 1840.

The patent laws of the United States, enacted, pursuant to the provisions of the Constitution, to promote the progress of the useful arts, secure to the originator of a new and useful invention, as a reward and encouragement for the exercise of his ingenuity, "the full and exclusive right and liberty of making, using, and vending to others to be used," for a limited period, the product of his genius and skill. This monopoly is expressly limited to a term of fourteen years, and at the end of that time, the invention becomes the property of the public and enures to its use. To protect the patentee in its exclusive enjoyment during the term of the monopoly, and at its expiration to insure to the public its beneficial use, the statute provides, that before the patent is issued, the inventor "shall deliver a written description of his invention 'or discovery, and of the manner and process of making, con- 'structing, using, and compounding the same, in such full, 'clear, and exact terms, avoiding unnecessary prolixity, as 'to enable any person skilled in the art or science to which 'it appertains, or with which it is most nearly connected, to 'make, construct, compound, and use the same; and in case 'of any machine, he shall fully explain the principle, and the 'several modes in which he has contemplated the application 'of that principle or character by which it may be distin- 'guished from other inventions." Another provision of the statute makes the patent void, if the specification "does not contain the whole truth relative to his invention," provided, "the concealment is for the purpose of deceiving the public." It follows, therefore, and such is the construction given to the statute, that the patentee is bound to describe, in his specification, fully and explicitly, the whole of his invention and its several parts, including every circumstance conducing to its advantageous use, with such accuracy as will prevent infringements during the term of the monopoly, and, at its close, enable the public, by the aid of skilful mechanics and competent workmen, without further invention, to put it into practice, and avail of its use.

"You know the object of the specification," says Tindal, C. J., in Neil *vs.* Harford, cited in Coryton on Patents, 42: "it is the price which the party who obtains the patent pays for it."

In Bovill *vs.* Moore, Dav. Pat. Ca. 400, the law on this point is clearly stated by Gibbs, C. J. "A patentee," says he, "who has invented a machine useful to the public, and 'can construct it in one way more extensive in its benefits 'than in another, and states in his specification only that 'mode which would be least beneficial, reserving to himself 'the more beneficial mode of practising it; although he will 'so far have answered the patent as to describe in his specifi- 'cation a machine to which the patent extends, yet he will 'not have satisfied the law by communicating to the public 'the most beneficial mode he was then possessed of for exer- 'cising the privilege granted to him." A wilful failure, therefore, to set forth and describe in the specification the whole of the invention as known and understood by the patentee, or a reservation of any essential part of it, or a concealment of the most beneficial way of constructing it, operates a fraud upon the public, and vitiates the patent. Nor can a valid patent be subsequently granted for that part of the invention which the patentee has concealed and failed to describe and claim in his specification, for that would enable him to take advantage of his own fraud, and, virtually, to secure a monopoly of twenty-eight, instead of fourteen years, for one invention.

If, then, the allegation of the remonstrants is true, it is clear that both of Morse's patents are void, and that the extension now sought cannot be granted. In support of the objection thus made, the testimony of Professor Leonard D. Gale, a witness produced on the part of the applicant, so far as it pertains to the present question, is cited and relied upon. Professor Gale is one of the counsel representing Morse in the pending application. He is, and has been, his intimate, bosom friend: during a long period, while Morse was engaged in experiments with a view to perfect his invention, he was his confidential assistant and adviser, and, for his services, was rewarded by an assignment of a considerable interest in the first patent. He, then, of all other men, excepting Morse himself, is supposed to have best understood and known the extent and scope of the invention patented in 1840; and, while it is asserted that his testimony is in conflict with sworn statements previously made by him in several suits at law and in equity involving the validity of these very patents, and the alleged discrepancy has, therefore, been visited with the most unsparing criticism, it is, nevertheless, confidently relied on as conclusive proof that the combination of the local circuit and local magnet was invented prior to the date of the patent of 1840.

That no further misconception as to the meaning of Professor Gale may exist, I shall quote his testimony as it was delivered from his own lips.

To the thirteenth interrogatory in chief, to-wit: "Do you know the date of the invention, by Morse, of local circuits?" he answers: "Mr. Morse first explained to me 'the use of the local circuit in the winter of 1837-8, in 'answer to questions which I had put to him previously 'concerning telegraphing through long lines." To the fourteenth interrogatory, to-wit: "Did he make himself completely understood to you at that time?" he answers: "He did." To the fifteenth interrogatory, to-wit: "From what he then said, could you then have constructed and used the local circuit as it is now used?" he answers: "I think I could; but no apparatus for that purpose had been made at that time." To the sixteenth interrogatory, to-wit: "Did he make any drawings to show the manner of constructing the local circuit?" he answers: "He did make a drawing, 'and also showed, by an arrangement of a link of iron and 'some wire, how he could arrange the local circuit in con'nection with the general circuit."

Such is the evidence relied upon by the remonstrants to establish the allegation that the invention of the combination of devices now in question was made as early as the winter of 1837-8, and formed a part and parcel of the system of telegraphing originally patented by Professor Morse.

Standing alone, and unqualified by its own express limitations, by other portions of the same deposition, and by the well known state of the invention at the date referred to, the testimony of Professor Gale is, perhaps, liable to some measure of misconception; but when these means of explanation are invoked in aid of its interpretation, it becomes manifest that the evidence of this witness has been perverted by an artful confusion of terms, and that it does not establish, nor was it intended to establish, the proposition for which the remonstrants contend. That the idea of the local circuit, as now used, is embraced in the principle of the combined or relay circuits, of Morse's first invention, may well be conceded; for, of that fact, I think there is but little doubt. But what was then understood by the term "local circuit," was a terminal circuit placed at the end of the line, constructed in the same manner, and holding the same relations to the successive links in the main line that they, consecutively, held to each other. The character and purpose of the magnet, the functions of the battery, the position of the register, were the same in both, and the only perceptible difference between them was in their relative length, the terminal being somewhat shorter than the relay circuit, in order that greater power might be imparted to the electric current for the purpose of working the register and recording the message at the end of the line. In this consisted the sole difference between them, and, as was well remarked in the argument, by one of the learned counsel, beyond this there was no invention; it was only experimental adaptation. The terminal circuit of the first patent was, and still is, with improved arrangements, known as the "local circuit."

Now, in the patent of 1846, the principle of the terminal circuit, with a modification and re-arrangement of the mechanism employed, has been applied to the lateral or office circuits.

The terminal circuit, made much shorter than before, and otherwise improved, while still employed at the end of the line, has been also located on the margin of the main circuit; the register has been removed from the main line to the termination of the short lateral or local circuit; the battery, which, under the old system, performed the double duty of working the register and breaking and closing the next succeeding circuit in the main line, is used to break and close the short, independent office circuit which works the register, while the relay magnet, which originally effected the combination of the circuits, has been substituted by the receiving magnet, which now connects the main and lateral or office circuits. This office or lateral circuit, thus arranged, is likewise known as the local circuit. It will thus be readily perceived how little of ingenuity is required to pervert, by confounding terms, the meaning of the most honest and truthful witness, when deposing in reference to the local circuit.

It will be remarked that the thirteenth interrogatory makes no distinction between these local circuits, nor does it designate to which of them it refers. It is general and undefined, and asks the witness for the date of the invention by Morse of "local circuits."

When Professor Gale comes to respond, he takes care so to answer as to qualify his meaning; and in stating that Morse first explained to him the use of the local circuit in the winter of 1837-8, he adds, that it was in answer to questions which he had put to him previously *concerning telegraphing through long lines.* This qualification is, of itself, sufficient to fix the character of the local circuit to which the witness referred, for it is hardly to be presumed that it would have been deemed necessary to elucidate the use of the local circuit in telegraphing through long lines, by explaining the arrangement for dropping messages at stations on the way. But Professor Gale does not stop here. As if apprehensive of misconception, and determined not to be misunderstood, he seizes the opportunity afforded by Mr. Eddy's first cross-interrogatory, to remove all doubt as to which of the two local circuits he meant to speak.

That cross-question was adroitly put, and well calculated to entrap the witness. It is as follows: "You say you have known the use of the combination of two circuits for twenty-two years, or since 1838; can it be new, April 11, 1846?" He answers, "I stated that Mr. Morse explained to me in the winter of 1837-8 the use of the local circuit; I should have said, in addition, *as adapted to long lines*." This answer clearly settles the difficulty, and removes all doubt as to which of the two local circuits the witness intended to testify. The combination and arrangement of the local circuits of the patent of 1846, had no reference whatever to the length of the line. The purpose of that combination was to deliver messages at way stations on the line. The number of these stations was determined solely by considerations of public convenience and business necessity, and whether the line was short or long, in no sense, affected its adaptation and arrangement. But, as has been already shown, the length of the line did materially affect the arrangement of the combined circuits of the patent of 1840, and particularly the terminal local circuit; for upon that arrangement depended the transmission of the electric current, and the power thereby imparted to it, in aid of its action upon the register. Again, in answer to a question, upon his re-direct examination, put by Charles Mason, Esquire, as follows: "You stated yester-'day that it was in the winter of 1837-8, that Morse first 'explained to you this combination of the local and general 'circuits; will you state whether there is any circumstance 'which enables you to fix that particular date, and, if so, 'what?" after fixing the date by the time when instruments were being made for the purpose of exhibiting the invention to Congress, he says "it was during the making of those instruments that Morse explained to me *how he would be able to carry messages through long lines.*"

All, then, that Morse ever explained to this witness in the winter of 1837-8, in regard to the local circuit, *was "concerning telegraphing through long lines;" "as adapted to long lines;" "how he would be able to carry messages through long lines;"* and as it has been already shown that the combination and arrangement of the local circuits of the patent of 1846, had no reference whatever to the length of the line, but that the combined circuits, including the local circuit of the patent of 1840, were expressly and only arranged for the purpose of transmitting messages through long lines, the irresistible conclusion is, that the testimony of Professor Gale referred to the latter, and not to the former combination.

But, assuming that this witness did mean to swear that Morse described to him, in the winter of 1837-8, the use of the local circuit of the patent of 1846, still his evidence is wholly insufficient to establish the invention at that period or at any other time before the patent of 1840 was issued.

To constitute invention, in the sense of the patent laws, the thing discovered must be perfected, and embodied in a distinct form, and carried into practical operation. Mere theory, or speculation, or experiment, never put into actual operation or tested by experience, does not amount to patentable invention. Tested by this well-settled rule of law, what, according to the evidence of Professor Gale, was the condition of the alleged invention of the local circuit at the period at which it is attempted to fix it? In answer to one of the interrogatories propounded to him, he says, that Morse, in his explanation of the use of the local circuit, made himself completely understood by him. In reply to another, he states that Morse made a drawing, and also showed, by an arrangement of a link of iron and some wire, how he could arrange the local circuit, in combination with the general circuit; but, when interrogated as to whether, from what Morse at that time said to him, he could then have constructed and used the local circuit in the way it is now used, he answers, doubtingly, "I think I could;" and then, positively, "but no apparatus for that purpose had been made at that time." And this is the entire evidence cited by the remonstrants to establish their allegation that the local circuit of the patent of 1846 was, in fact, invented as early as the winter of 1837-8.

That such a discovery as the witness has described, having no existence save in the mind, or in the verbal explanations of its pretended creator,—which was never seen, but in rough drawings, or in the crude form of a link of iron and some wire,—without body or shape,—with no apparatus constructed to test its capacity for practical use,—which the witness who testifies of it, never saw in operation, and from the description given of it only *thinks*, but does not *know*, that he could have constructed, is not an invention within the letter or spirit of the patent laws, nor entitled to the protection of letters patent, can scarcely be considered worthy of grave argument for the purpose of demonstration. And if Professor Gale really intended to prove that the combination of devices constituting the main claim of the patent of 1846, was invented by Morse in the winter of 1837-8, his testimony, squared by the law, as it has been above stated, is, in my judgment, wholly insufficient for the purpose, and fails utterly to establish the fact. Fortunately, however, this office is not left to depend upon the testimony, alone, of Professor Gale for the solution of the question involved in the objection now under consideration. As before stated, Morse's deposition in the suit of French *vs.* Rogers has been made, by consent, evidence in this case. As a witness, he is unimpeached and unimpeachable, and is, therefore, entitled to full credit. He fixes the date of the invention, beyond all question.

In answer to the twelfth cross-interrogatory on the part of the complainant, he says: "The phrase local circuit refers 'rather to the relative arrangement of circuits than to a cir- 'cuit having, in itself, special distinctive peculiarities. If by 'the question is sought the date of my invention of breaking 'and closing one circuit by another, I answer, in 1836. I 'exhibited the same in operation in the spring of 1837. If 'by the question is sought the date of my invention of a short 'circuit to be used at the extremities of the line, I answer, in 'May, 1844. If by the question is sought the date of my 'invention of a still greater improvement; to-wit, that of 'placing short circuits on the margin (so to speak) of the 'main line, all of them to be operated simultaneously, I 'answer, that the idea of such improvement first presented 'itself to my mind in the beginning of the year 1844." He further states, that the short circuit at the extremities of the line was first used in May, 1844, and, the lateral circuits, in 1846, on the line between Philadelphia and Baltimore. Thus, in my opinion, is effectually put to rest the objection to the validity of the patent of 1846, upon the ground that the combination therein embraced was invented in the winter of 1837-8.

But, it is further insisted, that if the combination of the local circuit and local magnet in the patent of 1846 was not embraced in the patent of 1840, and if that combination was not invented by Morse as early as 1837-8, it is included in, and covered by, the English patents of Cook and Wheatstone and of Edward Davy, enrolled, respectively, December 12th, 1837, and January 4th, 1839, and, since both of these patents have priority over the patent of 1846, the latter is void, and ought not to be extended. A brief reference to the specifications and claims of these patents, so far as they relate to the subject, will suffice to dispose of this objection.

From these, it may be gathered that the invention of Cook

and Wheatstone embraces a system for giving signals, and sounding alarms in distant places, by means of electric currents, transmitted through metallic circuits. The system is made up of many details, which constitute parts of the improvement patented; among which are, a transmitting station, connected by continuous wires with a receiving station, which may be one of a series extending to the terminus, and there connected with a galvanic battery, so as to propel a current of electricity through the entire circuit.

This complete circuit requires two separate lines of insulated wire, and may use any number of batteries at a similar number of stations, so long as all are connected with the main line. Thus, one wire is used for transmitting, and one for returning the signal.

These signals constitute a telegraphic language or mode ot communicating letters of the alphabet, and numerals, or symbolic characters, produced by mechanism put into operation by the electric current from the transmitting station. The current, thus created, gives the required signal through a series of stations, or from the transmitting to the terminating points, or to any intermediate point, so long as there is a battery at each station in the main line; and when the current is insufficient to carry the signal between the termini of the line, the batteries at the intermediate stations are relied upon to supply an independent, though auxiliary, current, for the purpose of renewing the impulse and, thereby, reaching the desired station. It is, I think, apparent, that the invention of Cook and Wheatstone did not contemplate, at all, the use of independent currents out of the main circuit, but connected with it by a lateral current having its own register, operated by a local battery. In this respect it resembled Morse's first invention, as messages could not be delivered by it, at way stations, without interruption to the transit of the current on the main line. It had, to be sure, its local circuit, placed, like Morse's, at the end of the line; but its functions were wholly unlike those of the local circuit of Morse's patent of 1840. Morse's local circuit was employed in recording the message. Cook and Wheatstone's telegraph operated by the deflection of a needle, fixed, vertically, upon a horizontal axis in front of a dial, upon which were engraved the letters of the alphabet. It made no permanent record, but communicated the message by evanescent signs. As the movement of the needle was noiseless, producing no sound, it was indispensable to the practical use of this system that the operator at the receiving station, should be notified when a message was about to be sent. The local circuit of this patent was employed, solely, for the purpose of giving such notice to the operator, being so arranged as to cause an alarm to be sounded by striking a bell when the battery was brought into action. From the above description of Cook and Wheatstone's invention, I think it manifest that it did not embrace the combination of Morse's patent of 1846.

Davy's patent involved the printing of the message or signal on a chemically prepared fabric; the mechanism for which was put into operation by an electric current at the transmitting station, by which an electro-magnet was operated by a local battery and magnets at the receiving station, one or more, at pleasure, consisting of coils of wire surrounding vibrating magnetic needles. When these coils were charged by the current of the battery at the transmitting station, or the battery of the main route, the needles were deflected by the action of the coils, and, coming into contact with the wires of the local battery, charged the electro-magnet, and operated an escapement through the armature of the electro-magnet.

The signs or marks upon the paper at the receiving station, were made by the current of the local battery when caused to act by the current of the main circuit. This very complex system required three wires, of which two formed the main circuit, in which the current was broken and closed at the transmitting station; and an independent battery, at the receiving station, was, by a connection of its circuit with the main circuit, brought into action to release a clock movement or weight that rotated a cylinder covered with a chemically prepared material, upon which the message was indicated by the contact of a series of electrical points.

Undoubtedly, in Davy's system, there was a local circuit, employed for recording, but it was placed at the extremity of the line, as were the local circuits of Cook and Wheatstone's invention and of Morse's first patent. Davy's patent does not refer to the employment of the local circuit except at the end of the line, and no provision, whatever, is even alluded to, in his descriptions and claims, for lateral currents out of, but connected with the main circuits. I have thus endeavored to review and consider, in its turn, every objection, at all pertinent to the subject, that has been urged against the validity of the patent granted to Samuel F. B. Morse on the 11th of April, 1846, and now sought to be extended; and I am clearly of opinion that the combination of devices embraced therein, is not to be found in any patent, or invention, or in any printed publication, or in public use, prior to its date; but that Morse is the original and first inventor thereof, that the patent granted therefor was properly issued, and is, therefore, valid and binding in law. In this conclusion I am glad to find myself fortified by the judgment of the Supreme Court of the United States in the case of O'Reilly et al. *vs.* Morse et al., before cited, in which, although several of the objections, now considered, were not raised, nor was the character of the invention as elaborately reviewed as I have found it necessary to treat it, yet, with the accustomed clearness of reasoning and force of argument of the learned Chief Justice, it was demonstrated and decided that the patent for the local circuits was properly granted, and valid.

The next subject to which the attention of this Office must be directed, is that of remuneration. Has Professor Morse received a reasonable and proper remuneration for the time, ingenuity, and expense bestowed upon his invention, and upon its introduction into use? It is insisted by the remonstrants, that, if he is the original and first inventor of the improvement covered by the patent, the remuneration received by him is ample and sufficient, and that, therefore, an extension for an additional term of seven years, ought not to be granted.

In obedience to the requirements of the statute, the applicant has filed a detailed statement of his receipts and expenditures; but by reason of the fact that many of the devices and arrangements of the patent of 1840, were found to be indispensable to the successful working of the improvement patented in 1846, the one being the basis of the other, the two inventions have been united from the earliest period at which Morse's system of telegraphing was put into operation.

As no standard could be established, not purely arbitrary, by which their relative value might be ascertained, it has proved impossible to adopt a rule for an apportionment of receipts and expenditures between the two patents. Hence, the joint receipts and expenses of the two inventions have been blended in the account. The statement is too much in detail to admit of a repetition of its items, and I shall, therefore, content myself with a reference to the general results under its different heads, as that is quite sufficient for the purposes of the present inquiry. It is, however, proper here to state, that the patent of 1840 was extended by this Office, in 1854, for the term of seven years from its expiration, and, in granting the extension, the then Commissioner thought it his duty to charge the entire net profits of the two inventions against that patent. For this reason, the applicant has so stated his account as to exhibit, separately, the income and expenses up to, and since, the date of the extension.

The aggregate results are as follows:

Receipts prior to extension of patent of 1840.

Stocks at present cash value	$154,795 00	
Cash for patent rights	38,979 00	
Cash for sales of stock	24,200 00	
Cash for dividends on stock	47,090 75	
		$265,064 75

Receipts since extension.

Stocks at eighty cents on par value	$35,666 68	
Cash for patent rights	15,177 56	
Cash for dividends on stock	63,170 05	
		114,014 29
Total		$379,079 04
Expenses prior to extension	$139,130 38	
Expenses since extension	21,139 17	
		160,269 55
Total net receipts from both patents		$218,809 49

It will be observed, that the stocks accounted for in the above statement, and received prior to the extension of the patent of 1840, are rated at their market value at the time this application was filed. This, the counsel of Morse contend, is erroneous, and they insist that the patent is chargeable only with their actual cash value at the time of their transfer. The evidence in the case clearly shows that these stocks were accepted by Morse as the price or consideration of an absolute and unconditional sale of so much of his interest in the invention as was then assigned, and that they were, at that time, worth but fifteen dollars on each share of one hundred dollars.

The applicant for an extension is required to furnish a statement in writing, under oath, of his receipts and expenditures, sufficiently in detail to exhibit a true and faithful account of profit and loss in any manner accruing to him from, and by reason of the invention.

It is the profit and loss *accruing from the invention,* and from no other source, for which the patentee is held to account. Now, had these stocks been converted into cash, at the time of their transfer, and the proceeds invested in the purchase of other property, neither an increase nor a decrease in its value, while in the hands of the patentee, resulting, in the one case, from his own management and thrift, or from a fortunate turn in the market, and in the other, from causes of an opposite character, would have constituted a legitimate charge of profit or loss in the statement of receipts and expenditures; for, in neither event, would the effect produced have accrued, in any manner, from the invention. If this view be correct, as I think it is, then no sufficient reason can be discovered why the appreciation, from similar causes, of the stocks themselves, if retained by the patentee, should be considered as the product of the invention, and treated as a proper item of charge in the account of profits accruing from the patent.

Such being the case, I am of opinion, that the sum which the stocks would have realized, if converted into cash at the time the sale of the patent right took place, is the amount at which they should be estimated, when regarded as remuneration accruing from the invention. The applicant has charged himself with the gross sum of $154,795.00, on account of these stocks, rating them at their present cash value, and he has credited $68,845.00 of that amount, as paid to Amos Kendall for his services as agent. Estimating their value, when transferred to Morse, at fifteen dollars on each share of one hundred dollars, he would be chargeable, in round numbers, with $88,000.00, of which, $39,000.00 should be credited as the proportion due to Kendall, and, on this basis, the statement ought to be corrected.

Again, it is contended that the two charges for dividends on stocks received before and since the extension of the first patent, have been erroneously brought into the statement, and that they should be stricken entirely from it. These dividends are the annual increment of the stocks accepted by Morse as the consideration of the sale and assignment of his interest in the patent, and, upon the principle that they are to be treated as cash, at their market value when transferred, it follows that the dividends declared upon them are no more chargeable against the patent, nor to be considered as part of the remuneration accruing from it, than if they were the income arising from other property purchased with the proceeds of the stocks. The dividends, thus accounted for, are not properly chargeable against the patent, and should be stricken from the account.

Correcting the statement in these particulars, it would stand as follows:

Stocks received before the extension	$88,000 00	
Cash for rights	38,979 00	
Cash for sales of stock	24,200 00	
Stocks received since extension at eighty cents of par	35,666 68	
Cash for rights	15,177 56	
		$202,023 24
Cr.		
Total expenses	$181,269 55	
Less excess in value of stock paid to Kendall	29,845 00	
		151,423 55
Net balance received by Morse		$50,599 69

It thus appears, that the net profit accruing from both inventions, is $50,599.69. It will be remembered that the entire net profit of the two inventions up to 1854, when the first patent was extended, was applied, by the then Commissioner, to the account of that patent. This was done upon the assumption that the improvement embraced in the patent of 1846 was of little or no value. Enough has been already said to explain, fully, the character of the improvement which that invention engrafted upon the telegraphic art, and to demonstrate, beyond all question, that if, when considered merely as the creation of inventive ingenuity, it may not be entitled to take rank with the first invention, in its practical effects, at least, it is possessed of equal, or perhaps, even of superior merit. Estimating the two inventions as of equal value, the net profit of $50,599.69 when divided between them, will give, as the proportion of each, $25,299.84. Is this a reasonable remuneration for the time, ingenuity, and expense bestowed upon the invention and its introduction into use? The aggregate value of the time, ingenuity, and expense of an inventor, is the criterion by which, under the law, this Office is to judge of the sufficiency of the remuneration received by him, in order to determine whether an extension of the patent should or should not be granted. The value of the time devoted to the discovery, perfection, and introduction into use of an invention, is readily reached by reference to the emolument accruing, or that might have accrued, to the inventor from the ordinary pursuits of life for which he may have fitted himself. Thus, the income of the professional man, the scholar, or the artizan, forms, generally, an unerring standard of the worth of his services, when, diverted from his professional calling, he undertakes to explore the uncertain, devious, and intricate pathway of invention. The expense incurred while the invention rests in experiment, and up to its completion and introduction into use, is a mere matter of account, ascertainable by any one of ordinary business capacity. But the worth of the ingenuity developed in the discovery, stands upon a different basis, and is determinable solely by the value and importance to the public, of the invention itself when completed and in operation. Although no criterion may exist by means of which this value can be, relatively, computed, there is, nevertheless, a standard in no sense arbitrary, by which it may be readily ascertained. That standard is found in the evidence of intelligent witnesses, familiar with the invention and its effects, who testify, among other things, as to the character and measure of benefits derived from it, the increased conveniences and business facilities it affords, the economy in time and money consequent upon its introduction, and the augmentation of the general wealth of the community flowing from its use.

In the present case, the value to the public of the invention in question, is established, as far as it is possible to estimate it in dollars and cents, by the testimony of witnesses experienced in its operations, and capable of forming a correct judgment in regard to it. One of these witnesses, Mr. Kendall, although he thinks the value of the telegraph can hardly be estimated in figures, is of opinion, that, embracing the two patents, it is not worth less to the public than six or seven million of dollars, and that that amount should be divided between them. He bases his opinion "upon the increased 'expedition with which business can be done by the business communities, by its importance in the intercourse between members of families and friends, and, particularly, as 'an instrument of the Government in the management of its 'affairs."

As an illustration of its value in one instance, he states, that the late Governor Marcy informed him, that if Morse's telegraph had been constructed to New Orleans during the war with Mexico, the Government would have saved, in a single day, a half a million of dollars. Another witness, Cyrus W. Field, when asked if the telegraph is not worth a hundred millions of dollars, replies, that its value cannot be estimated in money, and that its importance can only be appreciated by the consequences which would result from its withdrawal from public use.

Assuming Mr. Kendall's estimate of the value of the two inventions constituting Morse's magnetic telegraph to be correct, what fair-minded man will assert that an individual who has invested the public with the use of property worth three millions five hundred thousand dollars, is adequately remunerated by the receipt in return for it, of $25,299.84, a sum not equal to the one seventh part of a single year's interest upon its estimated value? Or where is to be found the person whose judgment and sense of justice are not warped by the instincts of self-interest, or by the baser and more ungenerous promptings of prejudice or of envy, who will undertake to maintain that $109,404.74, the moiety of the net receipts with which the applicant in this case has erroneously charged himself, and but a fraction over one half of the yearly income of three and a half millions of dollars, is a reasonable compensation for a discovery, equal in value to the last named sum, dedicated by its owner to the public use? In my opinion, neither of these sums, nor both together, constitute a reasonable remuneration, within in the meaning of the Constitution and the patent laws, for the time, ingenuity, and expense bestowed upon this invention and upon its introduction into use; and I have no hesitation, whatever, in announcing my entire concurrence in the views of one of the witnesses who testified on the subject, that a half a million of dollars is but a moderate compensation to its inventor for the inestimable benefits which his discovery has conferred upon his country and the world.

The sum of six or seven millions of dollars at which its value has been set by the evidence on file, every well informed man must know, is but a tithe of its real value; and it was a well merited, though, perhaps, an unpremeditated tribute to the genius of its gifted author, when it was declared by competent judges, that this extraordinary emanation of his inventive ingenuity, is incapable of computation by any standard which the coinage of the mint affords, and that its intrinsic importance can never be appreciated, unless, suddenly blotted from existence, the business of the country and of the civilized world is left to languish and stagnate by reason of its loss. If an event so disastrous to every department of commerce and trade, to the varied pursuits of productive labor and to all the public and private interests of the country, could possibly happen, I hazard little in asserting, that, in less than one month from its occurrence, a hundred millions of capital would burst from the coffers where enterprise or avarice may have stored it, to seek a profitable investment in the purchase of the telegraph and its permanent restoration to use.

One more objection to the extension of this patent remains to be briefly considered, before this protracted discussion can be brought to a close. It is alleged that the applicant has transferred to the American Telegraph Company, all his interest in the extended term of the patent, excepting in some unimportant portions of the United States, and that, if the extension is granted, it will enure, almost exclusively, to the benefit of the assignees. The only evidence relied on to establish this allegation, is the testimony of Cyrus W. Field, who, in answer to certain cross-interrogatories, states that he is under the impression that the American Telegraph Company agreed to give Morse thirty thousand dollars in stock, in case the patent was extended. This is, certainly, neither competent nor sufficient evidence to prove an assignment of any interest whatever, in a patent; for if a contingent right in the future extension of a patent, is, in law, the subject of an assignment at all, the evidence of such assignment is required to be in writing, and to be recorded, within a limited time, in the Patent Office. Being matter of record, parol evidence is incompetent to prove it, and it can only be established by the production of the record itself, or a certified copy of it. If such evidence existed, it was readily accessible, and it was the duty of the remonstrants to have produced it, and failing to do so, the allegation must be considered as unfounded, and cannot be sustained. But, if an assignment were ever made of any portion of Morse's future rights in the extension of the present patent, the testimony abundantly shows, that he has reserved a large interest in stocks of the American Telegraph Company, and of other telegraphic corporations using his invention, besides the absolute right to the extended patent in California, Florida, and Texas; and, undoubtedly, this is a sufficient interest to support and justify the grant of an extension.

Upon the whole, I am of opinion that Morse is the original and first inventor of the combination of devices embraced in the patent granted to him on the 11th of April, 1846, and re-issued on the 13th of June, 1848; that the invention was novel and patentable when the patent was issued; and that, without neglect or fault, on his part, he has failed to receive a reasonable remuneration for his time, ingenuity, and expense bestowed upon it, and upon its introduction into use, and is, therefore, entitled to an extension of his letters patent.

It is, thereupon, ordered that the said letters patent, be, and the same are hereby extended for the term of seven years, from and after the expiration thereof.

PHILIP F. THOMAS,
Commissioner of Patents.

www.ingramcontent.com/pod-product-compliance
Lightning Source LLC
LaVergne TN
LVHW011124110826
845150LV00008B/2250

* 9 7 8 1 4 1 8 1 9 4 9 3 2 *